EDWARDIAN FARM

EDWARDIAN FARM

RURAL LIFE AT THE TURN OF THE CENTURY

Alex Langlands, Ruth Goodman
& Peter Ginn

First published in the United Kingdom in 2010 by
Pavilion Books

An imprint of Anova Books Company Ltd
10 Southcombe Street
London W14 0RA

Produced in association with Lion Television Ltd,
26 Paddenswick Road, London W6 0UB

(See picture credits on page 281 for further
information.)

Associate Publisher: Anna Cheifetz
Project Editor: Katie Deane
Designer: Nichola Smith
Cover Designer: Georgina Hewitt
Production Manager: Laura Brodie
Location Photographer: Laura Rawlinson
Copy Editor: Sharon Amos
Proofreader: Kathy Steer
Indexer: Sandra Shotter

ISBN 978-1-86205-885-9

A CIP catalogue record for this book is available
from the British Library.

10 9 8 7 6 5 4 3 2 1

Reproduction by Mission, Hong Kong
Printed and bound by L.E.G.O. S.p.A, Italy

www.anovabooks.com

Contents

INTRODUCTION

Edwardian Farm is, of course, the sequel to the Victorian Farm adventure that we undertook with Ruth, Alex and Peter in 2008 and very much the result of *Victorian Farm's* huge success. The six-part TV series peaked with six million viewers a week on BBC2 – one of the channel's highest audiences of the year – and the book went to No.1 on the bestseller list.

When the DVD of *Victorian Farm* went straight to No.2 on the Amazon pre-sales list (just behind the latest James Bond movie, *Quantum of Solace*) we all knew something a bit special was going on. Overwhelmed by the popularity of what we had created, we knew we had to follow it up. But in what form? *Victorian Farm* traced a year in the life of a farm in Shropshire as it would have been around the year 1885. But would people have an appetite for yet another year on a farm? Should we look instead at a different aspect of life? Should we move on to a completely different era?

WHAT NEXT?

My strong feeling was that we needed to retain the magical charm we got from the Victorian era. And I felt there was still a great deal of mileage left in farming. The trick we needed to pull off was how to retain these two winning elements without repeating ourselves. Having not only spent a year making *Victorian Farm* but also following it up with a three-part *Victorian Farm Christmas* mini-series in 2009, there was a good deal of understandable scepticism about how we could possibly find new things to do.

The answer was deceptively simple. It lay in Britain's extraordinary regional diversity. For a tiny island, it's astonishing just how divided we are in terms of accents, dialects, customs and architecture – let alone the endless variations in cuisine, dress and clothing, and even tools in different parts of the country. It always stuck in my mind, for example, that Henry Stephens' *The Book of the Farm* (our farming bible on

A view of the location, Morwellham Quay, Devonshire.

Victorian Farm) found it necessary to devote two pages of illustrations to the humble pitch fork – revealing dozens of different varieties all forged with the same job in mind, but all differing according to the region they were made in. Lincolnshire, Lancashire, Yorkshire, Shropshire, Devonshire, Derbyshire, etc., all seem to have insisted on their own version of this simple tool. And that's just in England. Who knows what Mr Stephens might have discovered in Scotland, Wales and Northern Ireland?

Victorian Farm was located in Shropshire and we stuck very much to the Shropshire way of doing things, using tools, methods and even breeds of livestock that were particular to the area in the 19th century.

I felt sure that if we went to a different region of Britain, we could uncover a whole new way to spend a farming year. Our research into different regions of the country proved this to be the case. And we found the most significant differences of lifestyle existed in coastal regions. Places where climate and conditions dictated a different way of life, a different farming evolution.

LOCATION

We settled on a location called Morwellham Quay in Devonshire. This is a historic site brimming with period buildings and equipment as well as being at the centre of a rural community that was a bustling hub in the 19th century.

David joins the cast for their Edwardian seaside experience.

Besides a working quayside with its own ship, the site also contains a mine, a cooperage, a blacksmith's forge, a village shop, a pub, a chapel and cottages, as well as a farmyard, barns and stables – all restored to pristine turn-of-the-century condition and well stocked with period tools and equipment.

Nestled on the banks of the River Tamar, Morwellham was once one of the busiest ports in Britain. The keys to its success were adaptivity and diversification. And this offered us an abundance of new crafts and skills to tackle.

The farmers harnessed the region's unique climate – combined with its shipping and rail routes – to become leading suppliers of daffodils, cherries and strawberries.

Farmers here didn't just rely on crops and livestock. For generations they had taken advantage of the proximity of the sea to supplement their income – living the lives of 'fisherman-farmers' who combined life on the land with seasonal fishing. They were similarly adept at turning their hands to mining – the region being home to the largest known deposits of copper and arsenic in the world.

A series of bad harvests, combined with cheap foreign imports, created an agricultural depression in the late-19th century. The ever-resourceful farmers of the Tamar Valley exploited a new industry: market gardening.

And then there are the opportunities offered by the unique landscape of nearby Dartmoor. No wonder it's designated an Area of Outstanding Natural Beauty.

WHY EDWARDIAN?

The other change made was to move ahead in time. If anyone shared the charm and appeal of the Victorians, it was the Edwardians. By moving along twenty years – from the 1880s to the 1900s – we planted ourselves in an era which connects Victorian tradition with 20th century modernity. The idea of *Victorian Farm* had been inspired by my recollections of the stories my grandmother used to tell me about her childhood on a farm in Sturminster Newton, Dorset. Much of what she used to tell me was handed down from her parents and older siblings and I automatically bracketed it as Victorian. But she – and my grandfather who grew up on a farm in Bideford, Devon – were born in the early 1900s and I realize now that much of the experience I translated as Victorian was actually Edwardian. Like most people, I have a natural tendency to merge to the two eras, as though the Edwardian age is merely a footnote to the Victorian.

For some strange reason we don't give the Edwardian period it's due. We neglect the era between the death of Queen Victoria in 1901 and the outbreak of The Great War in 1914. Yet the short reign of Edward VII saw one of the most dynamic periods of social, economic and technological change in our history.

While most of the great innovations we associate with the modern world – gas, electricity, the combustion engine, etc. – were invented by the Victorians, it was the Edwardians who found practical ways to harness them for mass production and everyday use (for example, the motorized omnibus, the production-line car, the electrical generator, moving escalators on the London Underground). Edwardian inventions ranged from the first aeroplane to the tractor and the thermos flask. Suffragettes and trade unions were thoroughly Victorian, but it was in Edwardian Britain that 'Votes For Women' became a popular cause and workers got state pensions and unemployment benefit. It's an age defined by *The Hound of the Baskervilles* (incidentally set in Devon), *The Thirty-Nine Steps*, *The Railway Children* and the works of Beatrix Potter.

I hope *Edwardian Farm* sheds new light on this pivotal and distinctive era as well as providing the natural successor to *Victorian Farm.*

All of this was enabled by the tremendous faith and belief shown by our BBC Commissioning Executive, Emma Willis, BBC2's Head of Planning and Scheduling, Kate Mordaunt, and Controller of BBC2, Janice Hadlow. I'm delighted by the scale and ambition of the Edwardian Farm project in tackling not only a new era in a new location, but also a whole new range of ordinary people's lives.

DAVID UPSHAL

Executive Producer, Lion Television

Meet the team

ALEX

At first glance there seems little difference between the farming of the Edwardian period and that of the Victorian period. Victoria's reign spanned nearly sixty three years (June 1837 – January 1901) and in that time huge advances had been made in agriculture. Edward VII's reign covered a mere ten years by comparison and change in the farming world was much less marked in this time frame – certainly, there was little to differentiate between the practices of the last years of Victoria's reign and the first years of Edward's reign. In many ways historical terms such as 'Victorian' and 'Edwardian' don't do justice to the complex processes of change that can be observed from the middle years of the 19th century through to the beginning of the First World War. They present us with ill-fitting timescales and are 'periods' defined merely by the coronation and death of a monarch who, in most instances, had nothing to do with the processes themselves. Much of what we see happening in Edwardian farming has its origins in the 1870s and 1880s and likewise, those distinctly 'modern' elements of agriculture that have their genesis in the Edwardian age don't become broadly accepted until after the First World War.

Throughout my childhood summers spent on holiday in Devon I had marvelled at the historic landscape around me and now was my chance to get involved with it.

But look a little deeper and the changes are there. The Edwardian period can be seen as a bridge between what is regarded as the 'old' and the 'modern' world. The changes are subtle but significant. Innovations in agricultural science were moving so quickly that practices in the field were being left behind. A further characteristic of farming in the first ten years of the 20th century was the dawning realization that agriculture was never really going to play the role in Britain's economy that earlier agriculturalists of the 1850s and 1860s had thought it capable of. There was, therefore, a kind of sink-or-swim entrepreneurialism among the farming community with those that could adapt and diversify being the most successful.

A change in landscape for this series also reflected the new issues we would be able to explore in the Edwardian farm. Since the late-18th century, Devon had been famed for its agricultural practices. The terrain and climate of the county had defined mankind's relationship with the natural environment and since the earliest days Devonian farming had developed a unique character with its own traditions and customs. Our location in Devon, Morwellham near Tavistock, is one of the few regions that reflected the diverse entrepreneurial activities now being taken up. I was wildly enthused about engaging with a whole range of other rural activities and exploring the

working countryside beyond farming. Fishing, mining and market gardening all played an important role in the late-19th- and early 20th-century rural economy and I was looking forward to finding out how these industries shaped the communities and landscape of the world around them. Diversification into any one of these industries was the way in which money was to be made and, always up for a challenge, I relished the opportunity to try my hand at as many avenues of possible income to see if I could make it as an Edwardian farmer.

RUTH

Every generation re-invents itself; that which was new and scary only a few years before becomes old hat. Fashion moves on, technology moves on, but most important of all, ideas move on. The Edwardian generation looked back on the Victorian values with the usual mixture of nostalgia and slight superiority. Few Edwardians would have been pleased to be confused with people of the 1880s. While to our eyes a hundred and more years on many of the basic practicalities of life seem identical, people living through the Edwardian years were conscious of enormous change. Sometimes it was the technology they noticed: motor cars, telephones and electric lighting were hugely exciting to people seeing them or using them for the first time. You could see the world almost change before your eyes as they rapidly moved from rarities to everyday items, at least in the towns. What a huge impact the social changes of the day must have made for our Edwardian forebears, too. The 1907 Pensions Act lifted the threat of old age in the workhouse from so many working people.

Better education and different job opportunities were real possibilities for many young people of both sexes. Politically, arguments raged about votes for women, about socialism and reform. Despite its brevity the Edwardian era is a surprisingly exciting one in the history of Britain and the British people. At the start of this project our aim was to experience elements of an Edwardian rural life. I hoped to gain a little insight into what it might have been like to be in the midst of that life – to see it from the inside.

There would be a lot of very hard work of course, I was under no illusions there, most of the labour-saving devices of the era were available only to the wealthy and the benefits of running water were rare in the countryside. Making a living too was going to be a challenge, rural Edwardians often struggled to make ends meet and had to become inventive 'jacks of all trades'. I knew too that my experience was going to differ quite substantially from that of Peter and Alex's. Of all periods of history the Edwardian is the one in which women have perhaps the least involvement in agriculture. One after another the traditional areas of women's work on the farm had been taken over by the men and the machines. Men had always dominated arable farming, horses and sheep. The late Victorian period had seen men take over the dairy herds and now even the poultry yard became men's work. Market gardening still employed large amounts of female labour, but, even so, there would be fewer times when the three of us worked side by side. My involvement on the farm would become one more of bookkeeping, marketing the produce, tapping into the potential of the tourist trade and providing the domestic support. No small task, any of them. Having learnt so much over the years working on our Victorian and Stuart farms I knew that I was going to need to learn a whole lot more this time around. A new rural economy, a new region and a new era all awaited me. I couldn't wait.

PETER

We feel the effects of our ancestors, and our descendants will feel the effects of us. From the Big Bang until the Big Crunch, the thread of human existence is one continuous strand, so complex and incomprehensible that we just about manage to grasp the concept of our own lives. Historians try to make sense of this and in doing so divide this tapestry of life into a patchwork quilt of pigeon-holed events. The Roman occupation of Britain, the Tudor period, the Victorian era – the past is dissected into pieces that are associated with a known historical entity or event.

The Edwardian period is just such an example. Purists may state that it ended in 1910 with the death of the King, others say it went on until the start of the Great War, some argue it continued to the end of the war and some state it is merely a continuation of the Victorian period. Any way that one cuts it makes no difference to the lives of the people who lived through it, and by undertaking a project such as *Edwardian Farm,* I am given an opportunity to glimpse fleetingly those very same lives.

Edwardian Britain saw massive social change coupled with major advancements in technology, however, as this project is concerned with the everyday existence of regional rural folk much of our focus was on the landscape and the traditions of the Tamar Valley. Furthermore, the turn of the century is tantalizingly close and the Edwardian period probably straddles the boundary of living memory giving it a sort of *presque vu* quality.

The era has been cemented into the national consciousness through the popular literature of the time such as *Wind in the Willows*, the *Tale of Peter Rabbit*, and *Peter Pan* along with photographic, music and cinematic recordings giving it a romantic quality that is further heightened by the commencement and subsequent suffering of the First World War. However, as with our previous farming projects 'our mission (should we choose to accept it)' is to delve into the day-to-day lives of the Edwardian inhabitants of this part of Britain and through reading, research, oral histories and most importantly the very act of doing, try to come close to what it would have been like to be one of the many faceless yet vital figures that adorn the annals of time. So with a shave and a haircut and a new set of clothes (mainly provided by my good friend Alex) I stepped into Edwardian Britain and braced myself for a new set of challenges.

EDWARDIAN FARMING AND THE END OF AN ERA

On the surface, there appears to be little change from the decades preceding Edward VII's accession to the throne. One could still describe the countryside itself as being in a state of steady decay and, overall, there was a general decline in the role played by farming in the wider economy of the British Isles. Whereas in 1867–9 farming had accounted for around one-sixth of the national output, by 1890 that figure had dropped to a tenth and on the eve of the First World War, it had plummeted to one-fifteenth. A far smaller percentage of the population were being fed from home-grown produce in 1909–13 than had been in 1870. But scratch the surface and look a little deeper and there are the signatures of the beginnings of our modern farming practices. A way of life and a harmonious relationship with the countryside that had at one point seemed timeless was being 'lost' and a new dawn; a farming of the modern age was emerging in its place.

FARMHOUSE INCOMES

New tastes among the swelling ranks of the urban middle-classes also provided farmers with opportunities. Certain foodstuffs that were considered traditionally to be the preserve of the farmer's wife such as fruit, vegetables, eggs and poultry were becoming more than added income to the farmhouse kitty and viable enterprises in their own right. Potatoes, for example, were one of the biggest cash crops of the first decade of 20th century and large areas of Lincolnshire, Yorkshire, Lancashire and Cheshire were turned over to this new money-maker. As far as the Southwest was concerned, its time-honoured pastoral economy was becoming increasingly supplemented by more specialist products such as strawberries, flowers, early potatoes, clotted cream, eggs and poultry. In many areas of the south coast, too, the growing number of middle-class tourists offered a highly profitable seasonal market.

EDWARDIAN COUNTRYSIDE

The first decade of the 20th century was a period within which British agriculture and the countryside in general experienced mixed fortunes. It is, perhaps, difficult to measure the scale of change in the Edwardian countryside simply because so many processes that we see passing through the early years of the 20th century had began as early as the late 1870s. Furthermore, the First World War had such an enormous bearing on both global and national affairs that it is difficult to see where farming as an industry might have gone had it not been for the special circumstances of war. Since British agriculture had crawled out of the depression of the late 1870s and 1880s, very little had changed up to 1914. Many of the agricultural implements that had been produced during the golden age of the mid-19th century were found to be so durable that, fifty years on, they were still being used by sons and grandsons. This might be good testimony to design of the implements themselves but the danger was that they were governing practices and skills which by the 20th century had become outmoded.

CHANGING TIMES

Farmers were becoming 'growers' – or horticulturalists – and producing a remarkable range of fresh fruit and vegetables as well as the traditional meat, dairy and cereal crops. The problem was that each farmer was selling individually his own produce to an often distant market with the rates for each transaction and transportation being high. There was a very real opportunity, if growers in certain areas came together, to form co-operatives and sell collectively. This would benefit everyone involved. For the growers there would be a reduction in overall costs, guaranteeing good prices for produce at market and for the consumers the flow of produce to market could be regulated lessening the impacts of both gluts and shortages. Growers could also purchase as a collective reducing the costs of seeds, fertilizers and feedstuffs.

Relative to other industries in Britain, agriculture provided the most appalling living and working conditions only in the Edwardian period, better education, cheaper newspapers and improved communications within rural society allowed labourers to see just how poor their lot was. Cities and towns represented new opportunities, better housing and job security while an existence in the field and the furrow meant a life of being underpaid and underfed. Even the most elegant of innovations of the period played a part in the decay of the working village. Rider Haggard, in his survey of rural England conducted in 1901–2, famously commented on how the bicycle allowed men from the villages around Peterborough to seek employment outside the village and farms and find new opportunities at the brickworks. Many just 'upped-sticks' and left, and in places this exodus left only those without the initiative and tenacity to work profit from the land. Those who stayed continued to get what they could from the soil and their livestock and when that failed they did what all farmers do in such situations and sought out supplementary incomes from second jobs in other rural industries such as fishing and mining.

THE GREAT WAR

It was a war of poisonous gas, barbed wire, balloon observation points, submarines, starvation, illness and death. Rapidly advancing technology was combined with and pitted against outdated tactics resulting in a casualty toll far exceeding anything the world had seen before.

To an extent it is the Great War that secures the memories in the national consciousness of the endless summers and carefree attitudes that preceded it in the Edwardian period. However, it is also the First World War that alters the British cultural landscape of a rigid class system and a division of wealth within the last age of the English country house. The Great War is a beast that no human will ever be able to fully comprehend and to this day we feel its effects upon our lives.

OVERLEAF Loading the precious daffodil harvest at a local railway for buyers in cities 'up-country'.